Volcanoes

Illustrations: Janet Moneymaker
Design/Editing: Marjie Bassler

Copyright © 2022 by Rebecca Woodbury, Ph.D.

Volcanoes
ISBN 978-1-950415-35-9

Published by Gravitas Publications Inc.
Imprint: Real Science-4-Kids
www.gravitaspublications.com
www.realscience4kids.com

Volcanoes can be very exciting!

Wow!

But what is a volcano?

Scientists study volcanoes to find out.

Recall that Earth is made of layers.

Earth is made of different layers.

The outer layer is the **crust.**

The middle layer is the **mantle.**

The innermost layer is the **core.**

The **crust** is solid.

The **mantle** is solid on top and soft like peanut butter below.

The **core** is soft metal on the outside and solid metal in the middle.

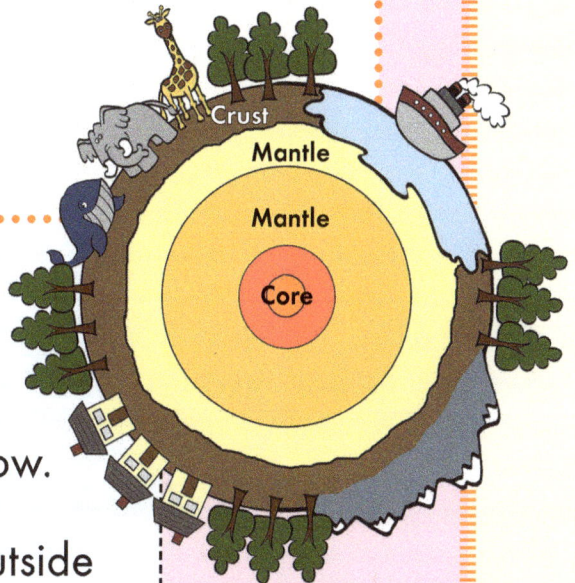

Crust

Mantle

Mantle

Core

Magma forms in the mantle when rocks and **minerals** deep in the Earth are squeezed. This squeezing causes the rocks and minerals to get so hot that they melt and become soft.

CRUST

MAGMA

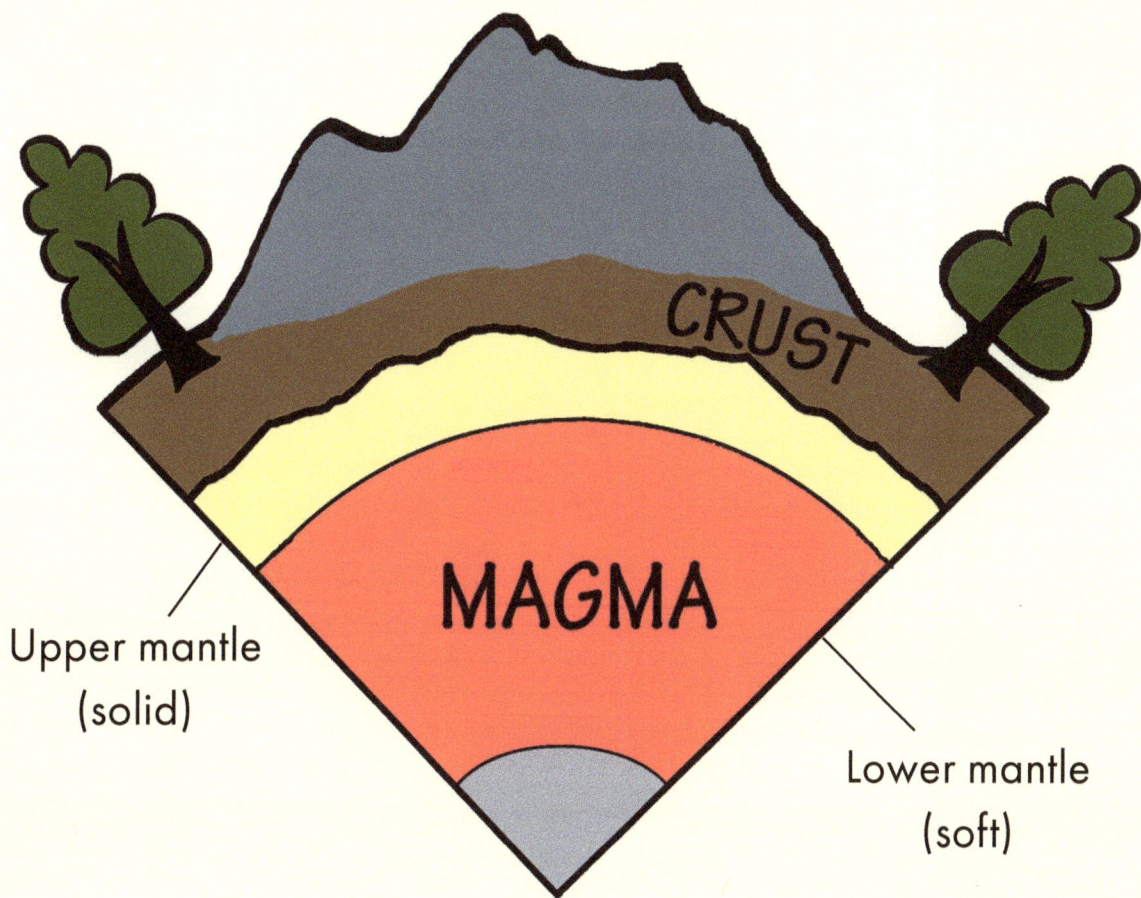

Upper mantle
(solid)

Lower mantle
(soft)

Volcanoes happen when hot magma inside the Earth pushes through a weak spot in the crust.

Magma that comes to the surface of Earth is called **lava.**

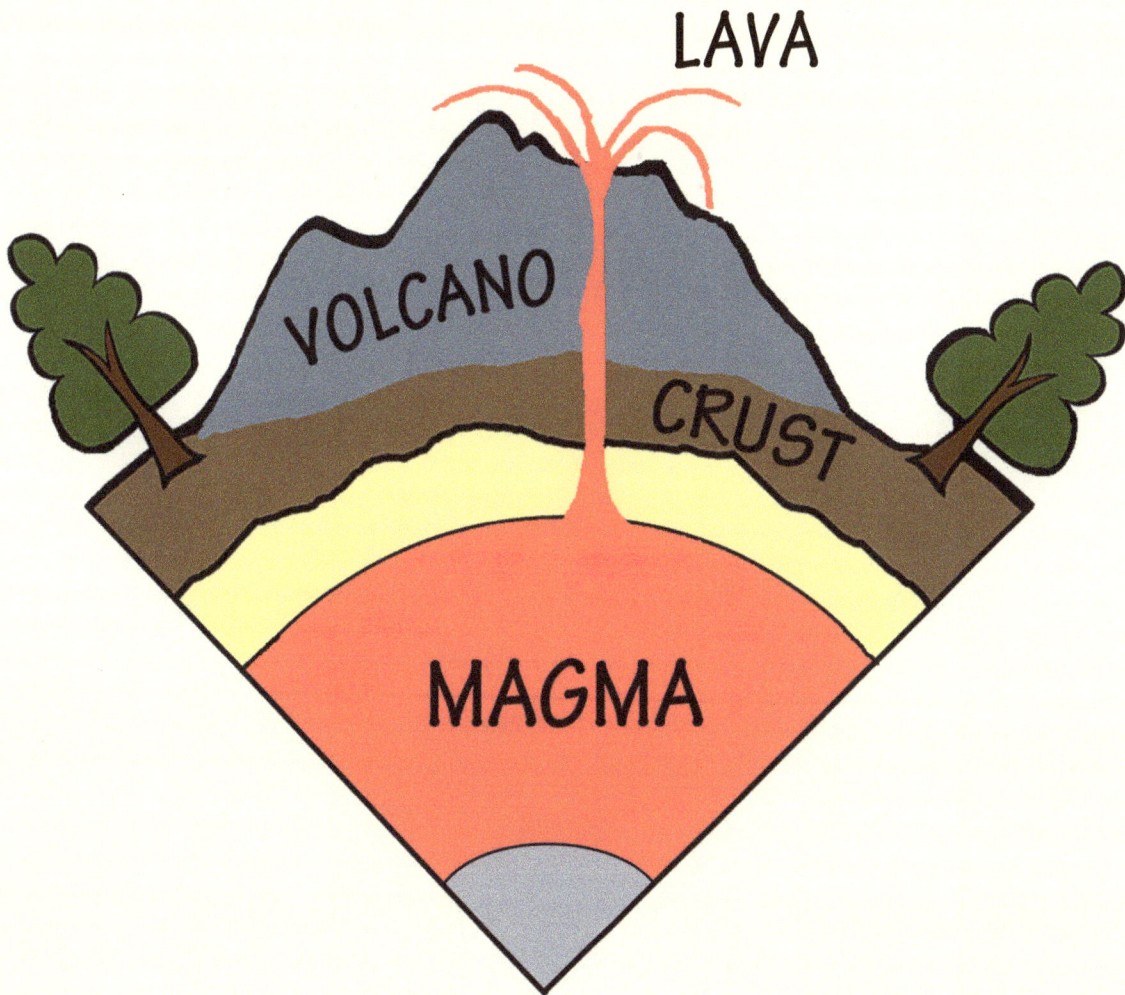

Lava can spout out of a volcano when the volcano **erupts.**

I didn't know rock could melt and do THAT!

Photo by Tanya Grypachevskaya on Unsplash

Lava can flow across land, making a thick river of hot melted rock.

Get out of the way!

Photo by Tanya Grypachevskaya on Unsplash

Mountains are made when lava cools and becomes hard rock.

Volcanoes make beautiful mountains!

Photo by ibrahim kusuma on Unsplash

Volcanoes can also create new islands in the ocean.

The **Hawaiian Islands** are the tops of volcanoes.

Those must be VERY TALL mountains.

They are!

We can learn many things about the inside of Earth by studying volcanoes, magma, and lava.

I want to learn more about volcanoes!

How to say science words

core   (KAWR)

crust   (KRUHST)

erupt   (i-RUHPT)

Hawaiian Islands   (huh-WIY-yuhn IY-luhndz)

lava   (LAH-vuh)

magma   (MAG-muh)

mantle   (MAN-tuhl)

mineral   (MIN-ruhl)

mountain   (MOWN-tuhn)

solid   (SAH-luhd)

volcano   (vahl-KAY-noh)

What questions do you have about VOLCANOES?

Learn More Real Science!

Complete science curricula from Real Science-4-Kids

Focus On Series

Unit study for elementary and middle school levels

Chemistry
Biology
Physics
Geology
Astronomy

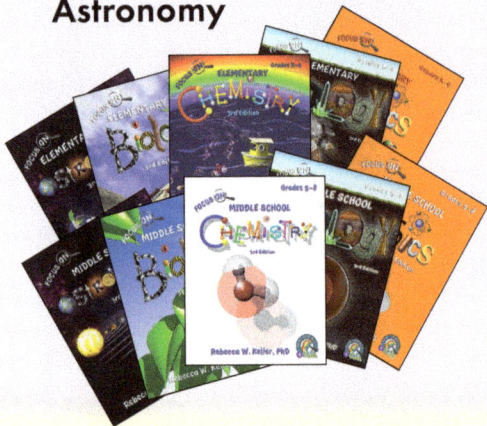

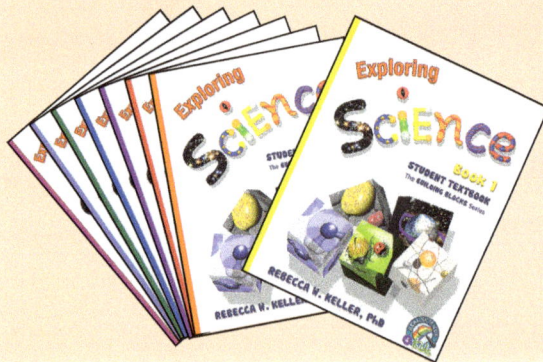

Exploring Science Series

Graded series for levels K–8. Each book contains 4 chapters of:

Chemistry
Biology
Physics
Geology
Astronomy

www.ingramcontent.com/pod-product-compliance
Lightning Source LLC
Chambersburg PA
CBHW040153200326
41520CB00028B/7589